Contents

1   Complex numbers

1.1   Arithmetic operations

Theorem. If $w = a + bi = r(\cos\theta + i\sin\theta)$ and $z = c + di = s(\cos\phi + i\sin\phi)$ are complex numbers, then:

1. $|wz| = |w||z|$;

2. $|w/z| = |w|/|z|$;

3. $\arg(wz) = \arg(w) + \arg(z)$;

4. $\arg(w/z) = \arg(w) - \arg(z)$.

Proof.

1.
$$\begin{aligned}
|wz|^2 &= |(a+bi)(c+di)|^2 \\
&= |(ac-bd) + (bc+ad)i|^2 \\
&= (ac-bd)^2 + (bc+ad)^2 \\
&= (ac)^2 - 2abcd + (bd)^2 + (bc)^2 + 2abcd + (ad)^2 \\
&= a^2(c^2+d^2) + b^2(c^2+d^2) \\
&= (a^2+b^2)(c^2+d^2) \\
&= |w|^2|z|^2 \\
&= (|w||z|)^2.
\end{aligned}$$

Since $|wz|$ and $|w||z|$ are both positive, we have that $|wz| = |w||z|$.

2.
$$\begin{aligned}
\left|\frac{w}{z}\right|^2 &= \left|\frac{a+bi}{c+di}\right|^2 \\
&= \left|\frac{(a+bi)(c-di)}{(c+di)(c-di)}\right|^2 \\
&= \frac{1}{(c^2+d^2)^2}|(a+bi)(c-di)|^2 \\
&= \frac{1}{(c^2+d^2)^2}(a^2+b^2)(c^2+d^2), \text{ by the above} \\
&= \frac{a^2+b^2}{c^2+d^2} \\
&= \frac{|w|^2}{|z|^2}.
\end{aligned}$$

Since $|w/z|$ and $|w|/|z|$ are both positive, we have that $|w/z| = |w|/|z|$.

3.

$$
\begin{aligned}
wz &= r(\cos\theta + i\sin\theta) \cdot s(\cos\phi + i\sin\phi) \\
&= rs(\cos\theta\cos\phi - \sin\theta\sin\phi + i(\cos\theta\sin\phi + \sin\theta\cos\phi)) \\
&= rs(\cos(\theta + \phi) + i\sin(\theta + \phi)).
\end{aligned}
$$

So $\arg(wz) = \theta + \phi = \arg(w) + \arg(z)$.

4.

$$
\begin{aligned}
\frac{w}{z} &= \frac{r(\cos\theta + i\sin\theta)}{s(\cos\phi + i\sin\phi)} \\
&= \frac{r(\cos\theta + i\sin\theta)(\cos\phi - i\sin\phi)}{s(\cos\phi + i\sin\phi)(\cos\phi - i\sin\phi)} \\
&= \frac{r((\cos\theta\cos\phi + \sin\theta\sin\phi) + i(\sin\theta\cos\phi - \cos\theta\sin\phi))}{s(\cos^2\phi + \sin^2\phi} \\
&= \frac{r(\cos(\theta - \phi) + i\sin(\theta - \phi))}{s \cdot 1} \\
&= \frac{r}{s}\left(\cos(\theta - \phi) + i\sin(\theta - \phi)\right).
\end{aligned}
$$

So $\arg(w/z) = \theta - \phi = \arg(w) - \arg(z)$.

$\square$

1.2   The exponential form of a complex number

Theorem. The following results hold for all real numbers θ:

1. $\cos\theta = (e^{i\theta} + e^{-i\theta})/2$;

2. $\sin\theta = (e^{i\theta} - e^{-i\theta})/2i$.

Proof.   1.

$$
\begin{aligned}
\frac{e^{i\theta} + e^{-i\theta}}{2} &= \frac{(\cos\theta + i\sin\theta) + (\cos(-\theta) + i\sin(-\theta))}{2} \\
&= \frac{\cos\theta + i\sin\theta + \cos\theta - i\sin\theta}{2} \\
&= \frac{2\cos\theta}{2} \\
&= \cos\theta.
\end{aligned}
$$

3

2.

$$\frac{e^{i\theta} - e^{-i\theta}}{2i} = \frac{(\cos\theta + i\sin\theta) - (\cos(-\theta) + i\sin(-\theta))}{2i}$$
$$= \frac{\cos\theta + i\sin\theta - \cos\theta + i\sin\theta}{2i}$$
$$= \frac{2i\sin\theta}{2i}$$
$$= \sin\theta.$$

$\square$

1.3   Solving polynomial equations

Theorem. If z_1 is a solution of $f(z) = 0$, where f is a polynomial function with real coefficients, then z_1^* is also a solution of the equation.

Proof. Let $f(z) = a_n z^n + a_{n-1} z^{n-1} + \ldots + a_1 z + a_0$.
    Then we have $a_n z_1^n + a_{n-1} z_1^{n-1} + \ldots + a_1 z_1 + a_0 = 0$.
    Taking the complex conjugate of both sides, we obtain

$$(a_n z_1^n + a_{n-1} z_1^{n-1} + \ldots + a_1 z_1 + a_0)^* = 0^*$$
$$(a_n z_1^n)^* + (a_{n-1} z_1^{n-1})^* + \ldots + (a_1 z_1)^* + a_0^* = 0^*$$
$$a_n (z_1^n)^* + a_{n-1} (z_1^{n-1})^* + \ldots + a_1 (z_1)^* + a_0 = 0$$
$$\text{since } a_i \in \mathbb{R} \text{ for all } i.$$
$$a_n (z_1^*)^n + a_{n-1} (z_1^*)^{n-1} + \ldots + a_1 z_1^* + a_0 = 0.$$

So z_1^* is also a solution of $f(z) = 0$. $\square$

1.4   De Moivre's theorem

Theorem. For any integer n and real number θ, we have

$$(\cos\theta + i\sin\theta)^n = \cos(n\theta) + i\sin(n\theta).$$

Proof. We use induction on n.

- **Base case:** $n = 0$. Then $(\cos\theta + i\sin\theta)^n = 1 = \cos 0 + i\sin 0 = \cos(n\theta) + i\sin(n\theta)$.

- **Inductive step:** Assume the result holds for $n = k$, so

$$(\cos\theta + i\sin\theta)^k = \cos(k\theta) + i\sin(k\theta).$$

We now prove that the result holds for $n = k + 1$:

$$
\begin{aligned}
(\cos\theta + i\sin\theta)^{k+1} &= (\cos\theta + i\sin\theta)^k \cdot (\cos\theta + i\sin\theta) \\
&= (\cos(k\theta) + i\sin(k\theta)) \cdot (\cos\theta + i\sin\theta) \\
&\text{(by the inductive hypothesis).} \\
&= \cos(k\theta)\cos\theta - \sin(k\theta)\sin\theta + i(\cos(k\theta)\sin\theta + \sin(k\theta)\cos\theta) \\
&= \cos(k\theta + \theta) + i\sin(k\theta + \theta) \\
&= \cos((k+1)\theta) + i\sin((k+1)\theta)
\end{aligned}
$$

So, by the Principle of Mathematical Induction, the results holds for all $n \in \mathbb{N}$.

We now wish to show that the result also holds for negative integers. Suppose n is such an integer, so $n < 0$. Then

$$
\begin{aligned}
(\cos\theta + i\sin\theta)^n &= \frac{1}{(\cos\theta + i\sin\theta)^{-n}} \\
&= \frac{1}{\cos(-n\theta) + i\sin(-n\theta)} \\
&\text{(by the above result).} \\
&= \frac{1}{\cos(n\theta) - i\sin(n\theta)} \\
&= \frac{\cos(n\theta) + i\sin(n\theta)}{(\cos(n\theta) - i\sin(n\theta))(\cos(n\theta) + i\sin(n\theta))} \\
&= \frac{\cos(n\theta) + i\sin(n\theta)}{\cos^2(n\theta) + \sin^2(n\theta)} \\
&= \cos(n\theta) + i\sin(n\theta) \\
&\text{(since } \sin^2(n\theta) + \cos^2(n\theta) = 1\text{).}
\end{aligned}
$$

Therefore the result holds for all $n \in \mathbb{Z}$. □

1.5 N^{th} roots of a complex number

Theorem. Suppose that $w = r(\cos\theta + i\sin\theta)$ is a nonzero complex number. Then the nth roots of w (the solutions of $z^n = w$ are the vertices of a regular n-gon centered at the origin in the complex plane.

Proof. For the nth roots to be vertices of a regular n-gon centered at the origin, they must:

1. Be equidistant from the origin, i.e. all lie on a circle with a certain radius.

2. Adjacent vertices have the same angle between them at the origin. As there are n vertices, adjacent vertices must form an angle of $2\pi/n$ at the origin.

Note that $z_0 = r^{1/n}(\cos(\theta/n) + i\sin(\theta/n))$ is one of the nth roots, since by de Moivre's theorem we get $z_0^n = r(\cos\theta + i\sin\theta)$.

In fact, consider $z_k = r^{1/n}(\cos((\theta+2\pi k)/n)+i\sin((\theta+2\pi k)/n))$. By de Moivre's theorem we get $z_k^n = r(\cos(\theta + 2\pi k) + i\sin(\theta + 2\pi k))$. This is equal to w for $k \in \mathbb{Z}$. So, taking $k \in \{0, 1, \ldots, n-1\}$ in z_k, we obtain all n roots of $z^n = w$.

All these roots have a modulus equal to $r^{1/n}$, and are therefore equidistant from the origin.

Furthermore, adjacent vertices are separated by an angle of $(2\pi(k+1))/n - 2\pi k/n = 2\pi/n$, and so they have the same angle between them.

Therefore, they are the vertices of a regular n-gon centered at the origin in the complex plane. $\qquad\square$

2   Matrices

2.1   Linear transformations

Theorem. Suppose M is the $n \times m$ matrix for a linear transformation $T : \mathbb{R}^m \to \mathbb{R}^n$. Then the ith column of M gives the new coordinates of the ith basis vector $\underline{e_i}$.

Proof. Note that the matrix M of a linear transformation is defined such that $M\underline{x}$ gives the new coordinates of $\underline{x}$. So consider $M\underline{e_i}$, where $\underline{e_i}$ has a 1 in the ith position and zeros elsewhere. Then

$$M\underline{e_i} = \begin{bmatrix} a_{11} & a_{12} & \cdots & a_{1m} \\ a_{21} & a_{22} & \cdots & a_{2m} \\ \vdots & \vdots & \ddots & \vdots \\ a_{n1} & a_{n2} & \cdots & a_{nm} \end{bmatrix} \underline{e_i}$$

$$= \begin{bmatrix} a_{1i} \\ a_{2i} \\ \vdots \\ a_{ni} \end{bmatrix}.$$

□

This is an important theorem as it allows us to construct the matrix for a given linear transformation simply by considering where the basis vectors go under the transformation.

Theorem. The matrix representing reflection of a point in $\mathbb{R}^2$ in the x-axis is given by

$$M = \begin{bmatrix} 1 & 0 \\ 0 & -1 \end{bmatrix}.$$

Proof. By the above theorem, we simply determine where the basis vectors go. $(1, 0)$ lies on the x-axis, so on reflection in the x-axis it is unchanged. $(0, 1)$, on the other hand, is reflected to $(0, -1)$. Thus the matrix for the transformation is $\begin{bmatrix} 1 & 0 \\ 0 & -1 \end{bmatrix}$.

□

Theorem. The matrix representing reflection of a point in $\mathbb{R}^2$ in the y-axis is given by

$$M = \begin{bmatrix} -1 & 0 \\ 0 & 1 \end{bmatrix}.$$

Proof. By a similar argument to above, $(1,0)$ is reflected to $(-1,0)$ and $(0,1)$ remains unchanged under the transformation. Thus the matrix is $\begin{vmatrix} -1 & 0 \\ 0 & 1 \end{vmatrix}$. □

Theorem. The matrix representing reflection of a point in $\mathbb{R}^2$ in the line $y = x$ is given by
$$M = \begin{bmatrix} 0 & 1 \\ 1 & 0 \end{bmatrix}.$$

Proof. The transformation sends $(1,0)$ to $(0,1)$ and vice versa.

□

Theorem. The matrix representing reflection of a point in $\mathbb{R}^2$ in the line $y = -x$ is given by
$$M = \begin{bmatrix} 0 & -1 \\ -1 & 0 \end{bmatrix}.$$

Proof. The transformation sends $(1,0)$ to $(0,-1)$ and $(0,1)$ to $(-1,0)$. □

Theorem. The matrix representing the anticlockwise rotation of a point in $\mathbb{R}^2$ through an angle θ about the origin is given by

$$M = \begin{bmatrix} \cos\theta & -\sin\theta \\ \sin\theta & \cos\theta \end{bmatrix}.$$

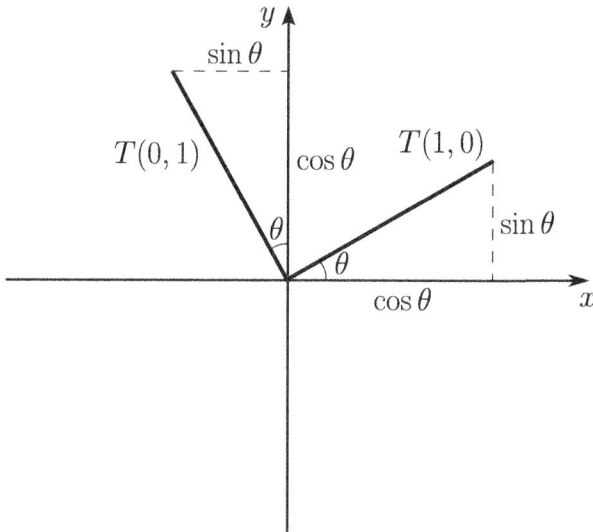

Figure 1: The rotation of the basis vectors $(1,0)$ and $(0,1)$ through an angle of θ radians anticlockwise about the origin.

Proof. Observe that, as shown in Figure 1, $(1,0)$ is transformed to $(\cos\theta, \sin\theta)$ and $(0,1)$ is transformed to $(-\sin\theta, \cos\theta)$. Therefore the matrix is $\begin{bmatrix} \cos\theta & -\sin\theta \\ \sin\theta & \cos\theta \end{bmatrix}$. □

Theorem. The matrix representing the stretch of a point in $\mathbb{R}^2$ by a scale factor k parallel to the x-axis is given by

$$M = \begin{bmatrix} k & 0 \\ 0 & 1 \end{bmatrix}.$$

Proof. $(1,0)$ is stretched to $(k,0)$ and $(0,1)$ remains unchanged as it has no component in the direction of the x-axis. □

Theorem. The matrix representing the stretch of a point in $\mathbb{R}^2$ by a scale factor k parallel to the y-axis is given by

$$M = \begin{bmatrix} 1 & 0 \\ 0 & k \end{bmatrix}.$$

Proof. $(1,0)$ remains unchanged as it has no component in the direction of the y-axis and $(0,1)$ is stretched to $(0,k)$. □

Theorem. The matrix representing the enlargement of a point in $\mathbb{R}^2$ by a scale factor k about the centre is given by

$$M = \begin{bmatrix} k & 0 \\ 0 & k \end{bmatrix}.$$

Proof. This time, $(1,0)$ and $(0,1)$ are both stretched to $(k,0)$ and $(0,k)$ respectively.

$\square$

We now progress to linear transformations in 3 dimensions. In 2 dimensions, the orientations of the axes are familiar and we can simply define a positive value of θ to be anticlockwise rotation. In 3 dimensions, the notion of orientation is a little less intuitive and whether a rotation is clockwise or anticlockwise depends on perspective. Thus, by convention, we assume that we are using a **right-handed coordinate system** as shown below:

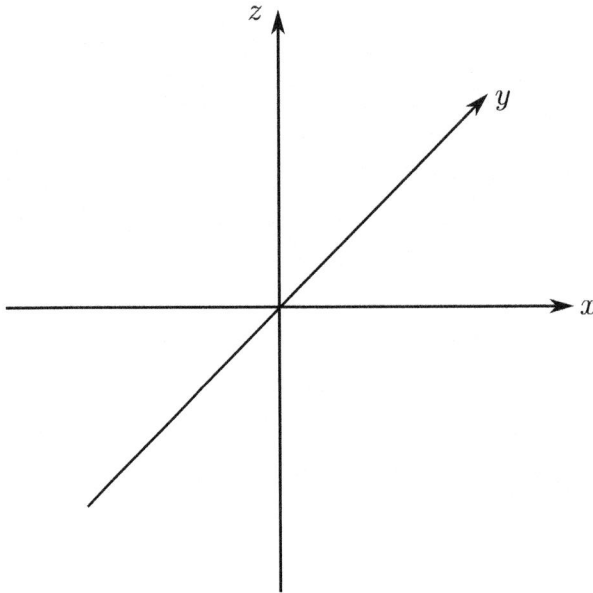

Figure 2: The coordinate axes in 3D.

This means that, on your right hand, your thumb, index finger and middle finger point in the directions of the x, y and z axes respectively.

We also assume that if we point our right thumb in the direction of the axis being rotated around, then the direction in which the fingers of your right hand curl gives the direction of rotation for positive values of θ.

Theorem. The matrix representing reflection of a point in $\mathbb{R}^3$ in the plane $x = 0$ is given by

$$M = \begin{bmatrix} -1 & 0 & 0 \\ 0 & 1 & 0 \\ 0 & 0 & 1 \end{bmatrix}.$$

Proof. $x = 0$ is the plane spanned by the y and z axes. Thus the basis vectors in this plane (i.e. $(0,1,0)$ and $(0,0,1)$) remain unchanged. $(1,0,0)$, on the other hand, is reflected to $(-1,0,0)$ on the opposite side of the yz-plane. □

Theorem. The matrix representing reflection of a point in $\mathbb{R}^3$ in the plane $y = 0$ is given by

$$M = \begin{bmatrix} 1 & 0 & 0 \\ 0 & -1 & 0 \\ 0 & 0 & 1 \end{bmatrix}.$$

Proof. Similar to the above, $y = 0$ is the xz-plane, thus $(1,0,0)$ and $(0,0,1)$ are unchanged but $(0,1,0)$ is reflected to $(0,-1,0)$. □

Theorem. The matrix representing reflection of a point in $\mathbb{R}^3$ in the plane $z = 0$ is given by

$$M = \begin{bmatrix} 1 & 0 & 0 \\ 0 & 1 & 0 \\ 0 & 0 & -1 \end{bmatrix}.$$

Proof. Similar to the above, $z = 0$ is the xy-plane, thus $(1,0,0)$ and $(0,1,0)$ are unchanged but $(0,0,1)$ is reflected to $(0,0,-1)$. □

Theorem. The matrix representing the anticlockwise rotation of a point in $\mathbb{R}^3$ about the x-axis is given by

$$M = \begin{bmatrix} 1 & 0 & 0 \\ 0 & \cos\theta & -\sin\theta \\ 0 & \sin\theta & \cos\theta \end{bmatrix}.$$

Proof. Clearly, the vector $(1,0,0)$ remains unchanged when rotated about the x-axis. Now we consider what happens to the other two basis vectors by orienting the coordinate system such that the x-axis goes into the page. The directions of the axes and the direction of rotation are given by the right-hand rules:

11

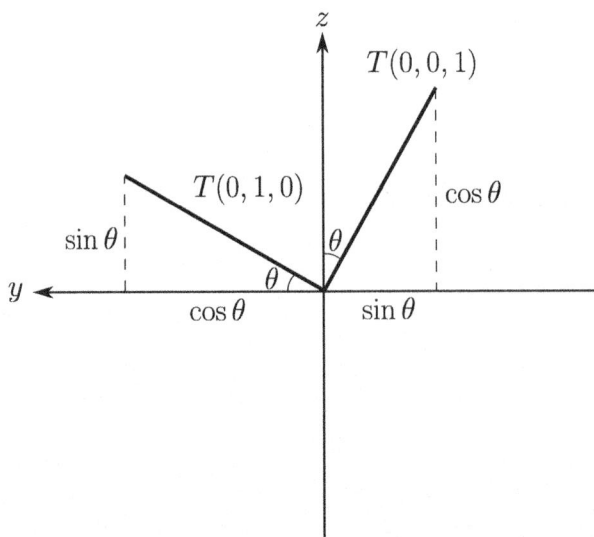

Figure 3: Rotation about the x-axis in 3D.

In Figure 3, we see that $(0, 1, 0)$ goes to $(0, \cos\theta, \sin\theta)$, since the x-coordinate is unchanged by the rotation. Similarly, $(0, 0, 1)$ goes to $(0, -\sin\theta, \cos\theta)$ (since $T(0, 0, 1)$ points away from the direction of the y-axis).

$\square$

Theorem. The matrix representing the anticlockwise rotation of a point in $\mathbb{R}^3$ about the y-axis is given by

$$M = \begin{bmatrix} \cos\theta & 0 & \sin\theta \\ 0 & 1 & 0 \\ -\sin\theta & 0 & \cos\theta \end{bmatrix}.$$

Proof. In a similar way to above, we orient the axes such that the y-axis goes into the page:

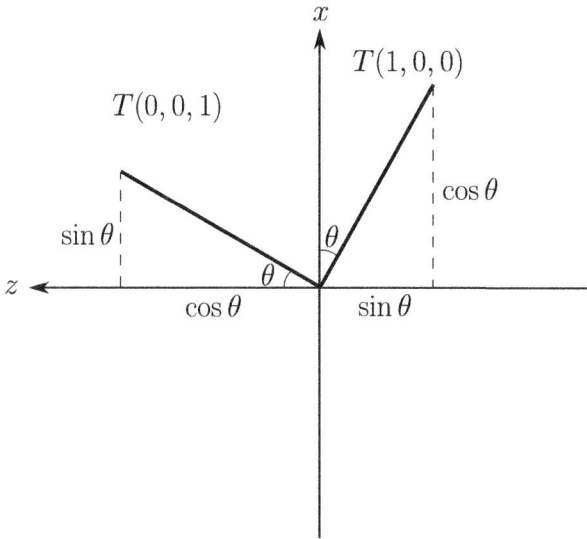

Figure 4: Rotation about the y-axis in 3D.

In Figure 4, we see that $(1, 0, 0)$ goes to $(\cos\theta, 0, -\sin\theta)$ and $(0, 0, 1)$ goes to $(\sin\theta, 0, \cos\theta)$.

$\square$

Theorem. The matrix representing the anticlockwise rotation of a point in $\mathbb{R}^3$ about the z-axis is given by

$$M = \begin{bmatrix} \cos\theta & -\sin\theta & 0 \\ \sin\theta & \cos\theta & 0 \\ 0 & 0 & 1 \end{bmatrix}.$$

Proof. Orienting the axes such that the z-axis goes into the page, we get the following:

13

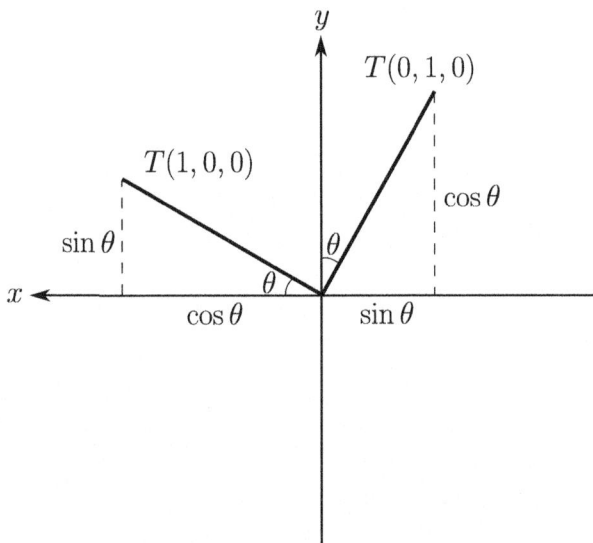

Figure 5: Rotation about the z-axis in 3D.

In Figure 5, we see that $(1,0,0)$ goes to $(\cos\theta, \sin\theta, 0)$ and $(0,1,0)$ goes to $(-\sin\theta, \cos\theta, 0)$. □

2.2 The inverse of a matrix

Theorem. A matrix M is invertible if and only if $\det M \neq 0$. Furthermore, if M is invertible, then its inverse M^{-1} is given by

$$M^{-1} = \frac{1}{\det M} \operatorname{adj}(M)$$

where adj(M) denotes the adjoint of matrix M.

Proof. Suppose that M is invertible. Then it has an inverse matrix, M^{-1}, such that $MM^{-1} = I$. Then $\det(MM^{-1}) = \det(I) = 1$. Since the determinant of a product of two matrices equals the product of their determinants, we get $\det(M)\det(M^{-1}) = 1$. Hence $\det(M) \neq 0$.

Conversely, suppose that $\det(M) \neq 0$. Now consider the matrix $M \operatorname{adj}(M)$. The ij^{th} entry of this matrix is the dot product of the ith row of M and the jth column of adj(M). Since adj(M) is the transpose of the matrix of cofactors of M, we get that the ij^{th} entry of $M \operatorname{adj}(M)$ equals

$$a_{i1}A_{j1} + a_{i2}A_{j2} + \ldots + a_{in}A_{jn}$$

14

where a_{ij} denotes the ij^{th} entry of M and A_{ij} denotes the ij^{th} cofactor of M.

If $i = j$, then this is the expansion of the determinant of M along the ith row, so it equals $\det(M)$.

If $i \neq j$, then this is the expansion of the determinant of a matrix similar to M, but with the jth row replaced by row i. Thus there are two copies of row i in this matrix, and so the determinant must be zero. This is because swapping the two rows should both negate the determinant and keep it the same (since the two copies of row i are identical, so swapping them doesn't change the matrix). This can only happen if the determinant is zero.

Hence we obtain

$$M \operatorname{adj}(M) = \begin{bmatrix} \det(M) & 0 & \ldots & 0 \\ 0 & \det(M) & \ldots & 0 \\ \vdots & \vdots & \ddots & \vdots \\ 0 & 0 & \ldots & \det(M) \end{bmatrix}$$

$$= \det(M)I$$

$$M\left(\frac{1}{\det(M)} \operatorname{adj}(M)\right) = I.$$

Therefore, by the definition of the matrix inverse

$$M^{-1} = \frac{1}{\det(M)} \operatorname{adj}(M).$$

$\square$

3   Further algebra

3.1   Roots of polynomials

Theorem. Suppose that the quadratic equation $ax^2 + bx + c = 0$ has solutions $x = \alpha$ and $x = \beta$. Then

1. $\alpha + \beta = -b/a$;

2. $\alpha\beta = c/a$.

Proof. By the Factor Theorem, we can write

$$ax^2 + bx + c = a(x - \alpha)(x - \beta)$$
$$= ax^2 - a(\alpha + \beta)x + a\alpha\beta.$$

Now comparing coefficients:

$$b = -a(\alpha + \beta) \qquad\qquad c = a\alpha\beta$$
$$\alpha + \beta = -\frac{b}{a} \qquad\qquad \alpha\beta = \frac{c}{a}$$

$\square$

Theorem. Suppose that the cubic equation $ax^3 + bx^2 + cx + d = 0$ has solutions $x = \alpha$, $x = \beta$ and $x = \gamma$. Then

1. $\alpha + \beta + \gamma = -b/a$;

2. $\alpha\beta + \alpha\gamma + \beta\gamma = c/a$;

3. $\alpha\beta\gamma = -d/a$.

Proof. As above, we use the Factor Theorem:

$$ax^3 + bx^2 + cx + d = a(x - \alpha)(x - \beta)(x - \gamma)$$
$$= a(x^2 - (\alpha + \beta)x + \alpha\beta)(x - \gamma)$$
$$= a(x^3 - \gamma x^2 - (\alpha + \beta)x^2 + (\alpha + \beta)\gamma x + \alpha\beta x - \alpha\beta\gamma)$$
$$= a(x^3 - (\alpha + \beta + \gamma)x^2 + (\alpha\beta + \alpha\gamma + \beta\gamma)x - \alpha\beta\gamma).$$

Again, comparing coefficients:

$$\alpha + \beta + \gamma = -\frac{b}{a}, \quad \alpha\beta + \alpha\gamma + \beta\gamma = \frac{c}{a}, \quad \alpha\beta\gamma = -\frac{d}{a}$$

$\square$

Theorem. Suppose that the quartic equation $ax^4 + bx^3 + cx^2 + dx + e = 0$ has solutions $x = \alpha$, $x = \beta$, $x = \gamma$ and $x = \delta$. Then

1. $\alpha + \beta + \gamma + \delta = -b/a$;

2. $\alpha\beta + \alpha\gamma + \alpha\delta + \beta\gamma + \beta\delta + \gamma\delta = c/a$;

3. $\alpha\beta\gamma + \alpha\gamma\delta + \beta\gamma\delta = -d/a$;

4. $\alpha\beta\gamma\delta = e/a$

Proof. Using the factor theorem

$$
\begin{aligned}
ax^4 + bx^3 + cx^2 + dx + e &= a(x - \alpha)(x - \beta)(x - \gamma)(x - \delta) \\
&= a(x^3 - (\alpha + \beta + \gamma)x^2 + (\alpha\beta + \alpha\gamma + \beta\gamma)x - \alpha\beta\gamma)(x - \delta) \\
&= a(x^4 - (\alpha + \beta + \gamma)x^3 + (\alpha\beta + \alpha\gamma + \beta\gamma)x^2 - \alpha\beta\gamma x + \\
&\quad (\alpha + \beta + \gamma)\delta x^2 - (\alpha\beta + \alpha\gamma + \beta\gamma)\delta x + \alpha\beta\gamma\delta) \\
&= a(x^4 - (\alpha + \beta + \gamma)x^3 + (\alpha\beta + \alpha\gamma + \alpha\delta + \beta\gamma + \beta\delta + \gamma\delta)x^2 \\
&\quad - (\alpha\beta\gamma + \alpha\beta\delta + \alpha\gamma\delta + \beta\gamma\delta)x + \alpha\beta\gamma\delta).
\end{aligned}
$$

Then comparing coefficients once again, we obtain the required results.   □

3.2   Common series

Theorem.

$$\sum_{r=1}^{n} r = \frac{1}{2}n(n + 1).$$

Proof. Let $S = \sum_{r=1}^{n} r$. Then by summing downwards from n instead of upwards from 1, we also get $S = \sum_{r=1}^{n}(n + 1 - r)$. Summing these two equations, we obtain

$$
\begin{aligned}
2S &= \sum_{r=1}^{n}(n + 1 - r + r) \\
&= \sum_{r=1}^{n}(n + 1) \\
&= n(n + 1) \\
S &= \frac{1}{2}n(n + 1).
\end{aligned}
$$

□

Theorem.

$$\sum_{r=1}^{n} r^2 = \frac{1}{6}n(n+1)(2n+1).$$

Proof. We prove this by induction on n:

- **Base case:** $n = 0$. Then $\sum_{r=1}^{n} r^2 = 0$ and $(1/6)n(n+1)(2n+1) = 0$.

- **Inductive step:** Assume that the result holds for $n = k$. Then we show that it also holds for $n = k + 1$:

$$\sum_{r=1}^{k+1} r^2 = \left(\sum_{r=1}^{k} r^2\right) + (k+1)^2$$

$$= \frac{1}{6}k(k+1)(2k+1) + (k+1)^2$$

(by the inductive hypothesis).

$$= \frac{1}{6}(k+1)[k(2k+1) + 6(k+1)]$$

$$= \frac{1}{6}(k+1)(2k^2 + k + 6k + 6)$$

$$= \frac{1}{6}(k+1)(2k^2 + 7k + 6)$$

$$= \frac{1}{6}(k+1)(k+2)(2k+3)$$

$$= \frac{1}{6}(k+1)((k+1)+1)(2(k+1)+1).$$

Therefore, by the Principle of Mathematical Induction, the result holds for all $n \in \mathbb{N}$. $\qquad\square$

Theorem.

$$\sum_{r=1}^{n} r^3 = \frac{1}{4}n^2(n+1)^2.$$

Proof. We again use induction on n:

- **Base case:** $n = 0$. Then $\sum_{r=1}^{n} r^3 = 0$ and $(1/4)n^2(n+1)^2 = 0$.

- **Inductive step:** Assume that the result holds for $n = k$. Then we show that it also holds for $n = k + 1$:

$$\sum_{r=1}^{k+1} r^3 = \left(\sum_{r=1}^{k} r^3\right) + (k+1)^3$$

$$= \frac{1}{4}k^2(k+1)^2 + (k+1)^3$$

$$= \frac{1}{4}(k+1)^2(k^2 + 4(k+1))$$

$$= \frac{1}{4}(k+1)^2(k^2 + 4k + 4)$$

$$= \frac{1}{4}(k+1)^2(k+2)^2$$

$$= \frac{1}{4}(k+1)^2((k+1)+1)^2.$$

Therefore, by the Principle of Mathematical Induction, the result holds for all $n \in \mathbb{N}$. □

3.3   Maclaurin series

Theorem. For all $x \in \mathbb{R}$, we have

$$e^x = \sum_{n=0}^{\infty} \frac{x^n}{n!}.$$

Proof. Let $f(x) = e^x$. Then $f'(x) = e^x$, and so it follows by induction that $f^{(n)}(x) = e^x$ for all $n \in \mathbb{N}$. Hence $f^{(n)}(0) = e^0 = 1$ for all $n \in \mathbb{N}$. So by the definition of a Maclaurin series, we have

$$e^x = \sum_{n=0}^{\infty} \frac{f^{(n)}(0)x^n}{n!}$$

$$= \sum_{n=0}^{\infty} \frac{x^n}{n!}.$$

□

Theorem. For all $x \in \mathbb{R}$, we have

$$\sin x = \sum_{n=0}^{\infty} \frac{(-1)^n x^{2n+1}}{(2n+1)!}.$$

19

Proof. Let $f(x) = \sin x$. We begin by calculating the first few derivatives:

$$f'(x) = \cos x$$
$$f''(x) = -\sin x$$
$$f'''(x) = -\cos x$$
$$f^{(4)}(x) = \sin x = f(x).$$

It follows by induction that

$$f^{(4k)}(0) = f(0) = \sin(0) = 0$$
$$f^{(4k+1)}(0) = f'(0) = \cos(0) = 1$$
$$f^{(4k+2)}(0) = f''(0) = -\sin(0) = 0$$
$$f^{(4k+3)}(0) = f'''(0) = -\cos(0) = -1$$

We can write the non-zero terms slightly differently, since $f^{(2(2k)+1)}(0) = f^{(4k+1)}(0)$ and $f^{(2(2k+1)+1)}(0) = f^{(4k+3)}(0)$.

Hence we can see that $f^{(2n+1)}(0) = 1$ if n is even and $f^{(2n+1)}(0) = -1$ if n is odd. Any other derivative of f that is not of this form equals zero when evaluated at zero, so other terms are not included in the Maclaurin series. Thus

$$\sin x = \sum_{n=0}^{\infty} \frac{(-1)^n x^{2n+1}}{(2n+1)!}.$$

$\square$

Theorem. For all $x \in \mathbb{R}$, we have

$$\cos x = \sum_{n=0}^{\infty} \frac{(-1)^n x^{2n}}{(2n)!}.$$

Proof. Let $f(x) = \cos x$. Taking the same approach as for $\sin x$, we get

$$f'(x) = -\sin x$$
$$f''(x) = -\cos x$$
$$f'''(x) = \sin x$$
$$f^{(4)}(x) = \cos x = f(x).$$

So by induction, we get $f^{(2(2k))}(0) = f^{(4k)}(0) = \cos 0 = 1$ and $f^{(2(2k+1))}(0) = f^{(4k+2)}(0) = -\cos 0 = -1$. The other derivative terms are zero, so the Maclaurin series is

$$\cos x = \sum_{n=0}^{\infty} \frac{(-1)^n x^{2n}}{(2n)!}.$$

$\square$

Theorem. For all $x \in (-1, 1)$, we have

$$\ln(1 + x) = \sum_{n=1}^{\infty} \frac{(-1)^{n+1} x^n}{n}.$$

Proof. Let $f(x) = \ln(1 + x)$. Then

$$f'(x) = \frac{1}{1 + x} = (1 + x)^{-1}$$
$$f''(x) = -(1 + x)^{-2}$$
$$f'''(x) = 2(1 + x)^{-3}$$
$$f^{(4)}(x) = -6(1 + x)^{-4}.$$

By induction it follows that $f^{(n)}(x) = (-1)^{n+1}(n-1)!(1+x)^{-n}$ for all $n \geq 1$. Hence $f^{(n)}(0) = (-1)^{n+1}(n-1)!$. So, in the Maclaurin series expansion of $f(x)$, we can rewrite each term $(f^{(n)}(0)x^n)/n!$ as $((-1)^{n+1}(n-1)!x^n)n! = ((-1)^{n+1}x^n)/n$.

Observe also that $f(0) = \ln 1 = 0$, so we can ignore the $n = 0$ term in the expansion. Therefore the Maclaurin series expansion is

$$\ln(1 + x) = \sum_{n=1}^{\infty} \frac{(-1)^{n+1} x^n}{n}.$$

$\square$

Theorem. For all $x \in (-1, 1)$ and $a \in \mathbb{R}$, we have

$$(1 + x)^a = \sum_{n=0}^{\infty} \binom{a}{n} x^n.$$

Proof. Let $f(x) = (1 + x)^a$. By the power rule and chain rule, we can see that $f^{(n)}(x) = a(a-1)\ldots(a-(n-1))(1+x)^{a-n}$. Thus $f^{(n)}(0) = a(a-1)\ldots(a-(n-1))$.

Now note that

$$\frac{f^{(n)}(0)}{n!} = \frac{a(a - 1)\ldots(a - (n - 1))}{n!}$$
$$= \frac{a!}{(a - n)!n!}$$
$$= \binom{a}{n}.$$

So $(f^{(n)}(0)x^n)/n!$ in the Maclaurin series expansion of $f(x)$ becomes $\binom{a}{n}x^n$. Therefore

$$(1 + x)^a = \sum_{n=0}^{\infty} \binom{a}{n} x^n.$$

$\square$

4   Further calculus

4.1   Volumes of revolution

Theorem. The volume V formed by rotating the graph of $y = f(x)$ through 2π radians about the x-axis between $x = a$ and $x = b$ is given by

$$V = \pi \int_a^b y^2 \, dx.$$

Proof. We begin by considering the volume formed by rotating the graph on a very small interval, say from $x = a$ to $x = a + \delta x$, and see that it approximates the volume of a thin cylinder:

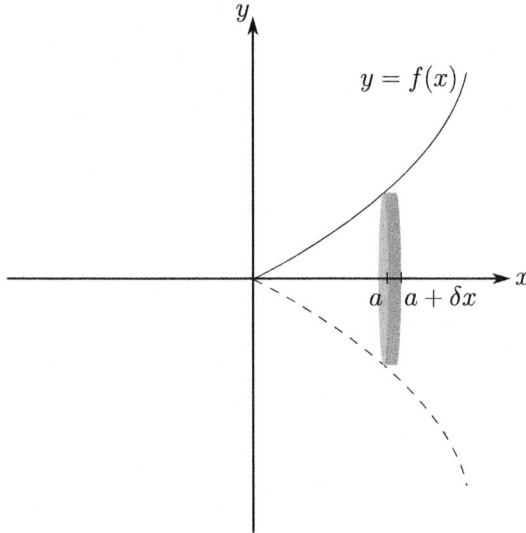

Figure 6: The volume of revolution formed by rotating the graph of $y = f(x)$ by 2π radians about the x-axis between $x = a$ and $x = a + \delta x$.

We know that the volume of a cylinder of radius r and height h is equal to $\pi r^2 h$. Thus, the volume of this thin cylinder is given by $\pi y^2 \cdot (a + \delta x - a) = \pi y^2 \cdot \delta x$. Suppose that we divide the interval from $x = a$ to $x = b$ into n intervals of equal size; each one then has a width of $\delta x = (b - a)/n$. And so the total volume is given by

$$V = \lim_{n \to \infty} \sum_{i=1}^{n} \pi y^2 \cdot \delta x$$

$$= \pi \lim_{n \to \infty} \sum_{i=1}^{n} y^2 \cdot \delta x$$

$$= \pi \int_a^b y^2 \, dx$$

$\square$

Theorem. The volume V formed by rotating the graph of $x = f(y)$ through 2π radians about the y-axis between $y = a$ and $y = b$ is given by

$$V = \pi \int_a^b x^2 \, dy.$$

Proof. The proof is symmetric, we have simply interchanged the roles of the x and y axes. Now, a cylinder positioned at a given position on the y-axis has a radius equal to the value of x for the function at that y-coordinate. We sum the volumes of these cylinders over the y-axis from $y = a$ to $y = b$, which is why we integrate with respect to y in the result. $\square$

4.2   Mean value of a function

Theorem. The mean value of a function $f(x)$ on the interval $[a, b]$ is given by

$$\frac{1}{b - a} \int_a^b f(x) \, dx.$$

Proof. The mean value of a function over an interval is loosely the 'average' value of that function over the interval. The average of a finite set of values is a familiar concept, but is trickier to define over an infinite set of values.

$$\lim_{n \to \infty} \frac{1}{n} \sum_{i=1}^{n} f(a + i\delta x) = \lim_{n \to \infty} \frac{1}{b - a} \sum_{i=1}^{n} f(a + i\delta x)\delta x$$

$$= \frac{1}{b - a} \lim_{n \to \infty} \sum_{i=1}^{n} f(a + i\delta x)\delta x$$

$$= \frac{1}{b - a} \int_a^b f(x) \, dx$$

$\square$

23

4.3   Derivatives of inverse trigonometric functions

Theorem. If $y = \arcsin x$, then

$$\frac{dy}{dx} = \frac{1}{\sqrt{1 - x^2}}.$$

Proof. We have that $x = \sin y$, so

$$\frac{dx}{dy} = \cos y$$

$$\frac{dy}{dx} = \frac{1}{\cos y}$$

$$\frac{dy}{dx} = \frac{1}{\pm\sqrt{1 - \sin^2 y}}$$

$$\frac{dy}{dx} = \frac{1}{\sqrt{1 - x^2}}.$$

Note that we take the positive square root since $\cos y$ is nonnegative for $-\pi/2 \leq y \leq \pi/2$, which is the domain of $\arcsin x$. □

Theorem. If $y = \arccos x$, then

$$\frac{dy}{dx} = -\frac{1}{\sqrt{1 - x^2}}.$$

Proof. We have that $x = \cos y$, so

$$\frac{dx}{dy} = -\sin y$$

$$\frac{dy}{dx} = -\frac{1}{\sin y}$$

$$\frac{dy}{dx} = -\frac{1}{\pm\sqrt{1 - \cos^2 y}}$$

$$\frac{dy}{dx} = -\frac{1}{\sqrt{1 - x^2}}.$$

Note that we take the positive square root since $\sin y$ is nonnegative for $0 \leq y \leq \pi$, which is the domain of $\arccos x$.

□

Theorem. If $y = \arctan x$, then

$$\frac{dy}{dx} = \frac{1}{1 + x^2}.$$

Proof. We have that $x = \tan y$, so

$$\frac{dx}{dy} = \sec^2 y$$

$$\frac{dy}{dx} = \frac{1}{\sec^2 y}$$

$$\frac{dy}{dx} = \frac{1}{1 + \tan^2 y}$$

$$\frac{dy}{dx} = \frac{1}{1 + x^2}.$$

$\square$

5   Further vectors

5.1   The dot product

Theorem. Let $\underline{a}$ and $\underline{b}$ be vectors in $\mathbb{R}^n$, and θ be the angle between them. Then $\underline{a} \cdot \underline{b} = |\underline{a}||\underline{b}| \cos \theta$.

Proof. Assume that $\underline{b}$ is a multiple of $\underline{a}$, say $\underline{b} = r\underline{a}$ for some $r \in \mathbb{R}$. Then $|\underline{a}||\underline{b}| \cos \theta = r|\underline{a}|^2 \cdot \cos(0) = r(\underline{a} \cdot \underline{a}) = \underline{a} \cdot (r\underline{a}) = \underline{a} \cdot \underline{b}$. So if $\underline{a}$ and $\underline{b}$ are multiples of one another, then $\underline{a} \cdot \underline{b} = |\underline{a}||\underline{b}| \cos \theta$.

Now we consider the case where $\underline{a}$ and $\underline{b}$ are not parallel:

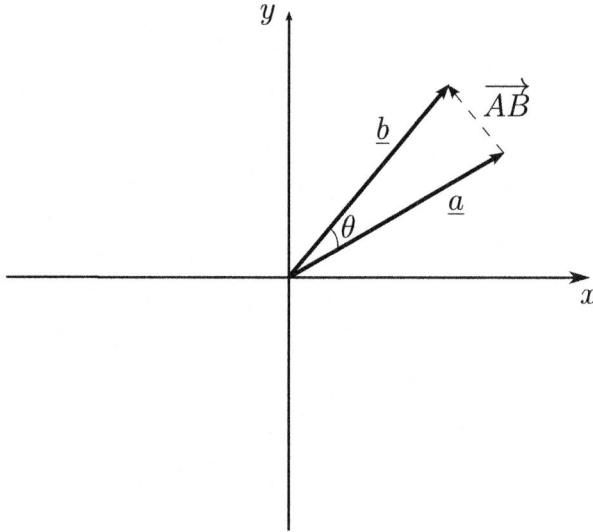

Figure 7: Two nonparallel vectors $\underline{a}$ and $\underline{b}$ with an angle θ between them.

Applying the cosine rule to Figure 7, we have

$$|\overrightarrow{AB}|^2 = |\underline{a}|^2 + |\underline{b}|^2 - 2|\underline{a}||\underline{b}| \cos \theta. \tag{1}$$

But we can rewrite $\overrightarrow{AB}$ as $\underline{b} - \underline{a}$, so $|\overrightarrow{AB}|^2 = (\underline{b} - \underline{a}) \cdot (\underline{b} - \underline{a})$. Substituting this into Equation 1, we get

$$(\underline{b} - \underline{a}) \cdot (\underline{b} - \underline{a}) = |\underline{a}|^2 + |\underline{b}|^2 - 2|\underline{a}||\underline{b}| \cos \theta$$
$$\underline{b} \cdot \underline{b} - (2\underline{a} \cdot \underline{b}) + \underline{a} \cdot \underline{a} = |\underline{a}|^2 + |\underline{b}|^2 - 2|\underline{a}||\underline{b}| \cos \theta$$
$$|\underline{b}|^2 - 2(\underline{a} \cdot \underline{b}) + |\underline{a}|^2 = |\underline{a}|^2 + |\underline{b}|^2 - 2|\underline{a}||\underline{b}| \cos \theta$$
$$-2(\underline{a} \cdot \underline{b}) = -2|\underline{a}||\underline{b}| \cos \theta$$
$$\underline{a} \cdot \underline{b} = |\underline{a}||\underline{b}| \cos \theta.$$

$\square$

Theorem. Two nonzero vectors $\underline{a}$ and $\underline{b}$ are perpendicular if and only if $\underline{a} \cdot \underline{b} = 0$.

Proof. Suppose that the two vectors are perpendicular. Then there is an angle of $\pi/2$ radians between them. Hence $|\underline{a}||\underline{b}| \cos \theta = |\underline{a}||\underline{b}| \cos(\pi/2) = |\underline{a}||\underline{b}|(0) = 0$.

Conversely, suppose that the dot product of the two vectors is 0. Using the above formula, we have $|\underline{a}||\underline{b}| \cos \theta = 0$ where $|\underline{a}| \neq 0$ and $|\underline{b}| \neq 0$. Therefore we must have $\cos \theta = 0$. There are then two possibilities: either $\theta = \pi/2$ or $\theta = 3\pi/2$. In either case, the two vectors are separated by an angle of $\pi/2$ radians. That is, they are perpendicular.

$\square$

5.2   Straight lines in 3D

Theorem. The Cartesian form of the equation of a straight line in 3D is

$$\frac{x - a_1}{b_1} = \frac{y - a_2}{b_2} = \frac{z - a_3}{b_3}$$

where a_1, a_2, a_3 are the components of $\underline{a}$ and b_1, b_2, b_3 are the components of $\underline{b}$.

Proof. The vector form of the equation of a straight line is $\underline{r} = \underline{a} + \lambda\underline{b}$, where $\underline{r} = (x, y, z)$. Let $\underline{a} = (a_1, a_2, a_3)$ and $\underline{b} = (b_1, b_2, b_3)$. Then

$$\begin{bmatrix} x - a_1 \\ y - a_2 \\ z - a_3 \end{bmatrix} = \lambda \begin{bmatrix} b_1 \\ b_2 \\ b_3 \end{bmatrix}.$$

Now, comparing components

$$\lambda = \frac{x - a_1}{b_1} = \frac{y - a_2}{b_2} = \frac{z - a_3}{b_3}.$$

$\square$

5.3   Planes in 3D

Theorem. The normal form of the equation of a plane in 3D is

$$\underline{r} \cdot \underline{n} = d$$

for some constant d and where $\underline{r} = (x, y, z)$.

Proof. Suppose we know that some point A lies on the plane. Then another point R lies on the plane if and only if the vector from A to R is parallel to the plane, that is perpendicular to the normal vector $\underline{n}$ for the plane. Let the position vector of A be $\underline{a}$ and the position vector of R be $\underline{r}$. Then

$$(\underline{r} - \underline{a}) \cdot \underline{n} = 0$$
$$\underline{r} \cdot \underline{n} - \underline{a} \cdot \underline{n} = 0$$
$$\underline{r} \cdot \underline{n} = \underline{a} \cdot \underline{n}$$
$$\underline{r} \cdot \underline{n} = d$$

where $d = \underline{a} \cdot \underline{n}$ is a constant. $\square$

Theorem. The Cartesian form of the equation of a plane in 3D is

$$ax + by + cz = d$$

where $a\hat{i} + b\hat{j} + c\hat{k}$ is a normal vector and d is a constant.

Proof. Suppose that $\underline{n} = (a, b, c)$ in the normal form of the equation of the plane. Then

$$\underline{r} \cdot \underline{n} = d$$
$$(x, y, z) \cdot (a, b, c) = d$$
$$ax + by + cz = d.$$

$\square$

Theorem. The distance s from a point (α, β, γ) to a plane given by the normal equation $\underline{r} \cdot \underline{n} = d$ is given by

$$s = \frac{|\underline{n} \cdot (\alpha\hat{i} + \beta\hat{j} + \gamma\hat{k}) - d|}{|\underline{n}|}.$$

Proof. Expanding the dot product in the result, we get

$$\frac{\underline{n} \cdot (\alpha, \beta, \gamma)}{|\underline{n}|} = \frac{|\underline{n}||(\alpha, \beta, \gamma)| \cos \theta}{|\underline{n}|}$$
$$= |(\alpha, \beta, \gamma)| \cos \theta.$$

$\square$

6   Polar coordinates

Theorem. The area A enclosed by a polar curve $r = f(\theta)$ between $\theta = \alpha$ and $\theta = \beta$ is given by

$$A = \frac{1}{2} \int_\alpha^\beta r^2 \, d\theta.$$

Proof. Imagine that the polar curve looks something like the following:

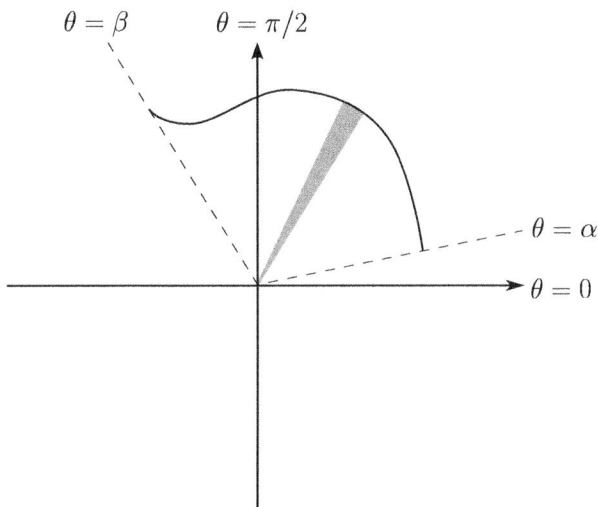

Figure 8: A polar curve $r = f(\theta)$, drawn between $\theta = \alpha$ and $\theta = \beta$.

Then the area of the small shaded section approximates that of a sector of a circle. This circle has a radius equal to the value of r at that point on the curve, and the circle has an angle of, say, $\delta\theta$. Thus the area of this sector is $(1/2)r^2\delta\theta$.

Let us vary n, and set $\delta\theta = (\beta - \alpha)/n$. Then

$$A = \lim_{n\to\infty} \sum_{i=1}^n \frac{1}{2} r^2 \delta\theta$$

$$= \frac{1}{2} \lim_{n\to\infty} \sum_{i=1}^n r^2 \delta\theta$$

$$= \frac{1}{2} \int_\alpha^\beta r^2 \, d\theta.$$

$\square$

7 Hyperbolic functions

7.1 Derivatives of hyperbolic functions

Theorem. The derivative of $\sinh x$ is $\cosh x$.

Proof.

$$
\begin{aligned}
\frac{d}{dx}(\sinh x) &= \frac{d}{dx}\left(\frac{e^x - e^{-x}}{2}\right) \\
&= \frac{e^x + e^{-x}}{2} \\
&= \cosh x.
\end{aligned}
$$

$\square$

Theorem. The derivative of $\cosh x$ is $\sinh x$.

Proof.

$$
\begin{aligned}
\frac{d}{dx}(\cosh x) &= \frac{d}{dx}\left(\frac{e^x + e^{-x}}{2}\right) \\
&= \frac{e^x - e^{-x}}{2} \\
&= \sinh x.
\end{aligned}
$$

$\square$

Theorem. The derivative of $\tanh x$ is $\operatorname{sech}^2 x$.

Proof.

$$
\begin{aligned}
\frac{d}{dx}(\tanh x) &= \frac{d}{dx}\left(\frac{\sinh x}{\cosh x}\right) \\
&= \frac{\cosh x \cdot \frac{d}{dx}(\sinh x) - \sinh x \cdot \frac{d}{dx}(\cosh x)}{\cosh^2 x} \\
&= \frac{\cosh^2 x - \sinh^2 x}{\cosh^2 x} \\
&= \frac{1}{\cosh^2 x} \\
&= \operatorname{sech}^2 x.
\end{aligned}
$$

$\square$

7.2   Integrals of hyperbolic functions

Theorem. The integral of $\sinh x$ is $\cosh x + C$, where C is a constant.

Proof. This follows from the Second Fundamental Theorem of Calculus.   □

Theorem. The integral of $\cosh x$ is $\sinh x + C$, where C is a constant.

Proof. This follows from the Second Fundamental Theorem of Calculus.   □

Theorem. The integral of $\tanh x$ is $\ln(\cosh x) + C$, where C is a constant.

Proof. Let $u = \cosh x$, so $du/dx = \sinh x$. Then we can use integration by substitution:

$$\int \tanh x \, dx = \int \frac{\sinh x}{\cosh x} \, dx$$
$$= \int \frac{1}{u} \, du$$
$$= \ln|u| + C$$
$$= \ln|\cosh x| + C$$
$$= \ln(\cosh x) + C.$$

Note that we removed the absolute value sign since $\cosh x \geq 1$ for all x.   □

7.3   Inverse hyperbolic function formulae

Theorem. The inverse of $\sinh$ is given by the formula

$$\operatorname{arsinh} x = \ln(x + \sqrt{x^2 + 1}).$$

Proof. Let $y = \operatorname{arsinh} x$, so $x = \sinh y$. Then

$$x = \frac{e^y - e^{-y}}{2}$$
$$2x = e^y - e^{-y}$$
$$2xe^y = e^{2y} - 1$$
$$e^{2y} - 2xe^y - 1 = 0.$$

Then, using the quadratic formula

$$e^y = \frac{2x \pm \sqrt{4x^2 - 4(1)(-1)}}{2(1)}$$

$$= \frac{2x \pm \sqrt{4(x^2 + 1)}}{2}$$

$$= x \pm \sqrt{x^2 + 1}.$$

But $e^y > 0$ for all y, so since $\sqrt{x^2 + 1} > x$ we must have that $e^y = x + \sqrt{x^2 + 1}$. It follows that $\operatorname{arsinh} x = \ln(x + \sqrt{x^2 + 1})$. $\qquad\square$

Theorem. The inverse of cosh is given by the formula

$$\operatorname{arcosh} x = \ln(x + \sqrt{x^2 - 1}).$$

Proof. Let $y = \operatorname{arcosh} x$, so $x = \cosh y$. Then

$$x = \frac{e^y + e^{-y}}{2}$$

$$2x = e^y + e^{-y}$$

$$2xe^y = e^{2y} + 1$$

$$e^{2y} - 2xe^y + 1 = 0.$$

Then, using the quadratic formula

$$e^y = \frac{2x \pm \sqrt{4x^2 - 4(1)(1)}}{2(1)}$$

$$= \frac{2x \pm 2\sqrt{x^2 - 1}}{2}$$

$$= x \pm \sqrt{x^2 - 1}.$$

Note that $y \geq 0$ since this is the range of $\operatorname{arcosh} x$, so $e^y \geq 1$ for all y. Hence $e^y = x - \sqrt{x^2 - 1}$ cannot be a solution, since for $x = 2$ we would have $e^y = 2 - \sqrt{3} < 1$. Thus we must have $e^y = x + \sqrt{x^2 - 1}$ and therefore $y = \ln(x + \sqrt{x^2 - 1})$. $\qquad\square$

Theorem. The inverse of tanh is given by the formula

$$\operatorname{artanh} x = \frac{1}{2} \ln\left(\frac{1 + x}{1 - x}\right).$$

32

Proof. Let $y = \operatorname{artanh} x$, so $x = \tanh y$. Then

$$x = \frac{e^y - e^{-y}}{e^y + e^{-y}}$$

$$x(e^y + e^{-y}) = e^y - e^{-y}$$

$$xe^y + xe^{-y} = e^y - e^{-y}$$

$$(1 - x)e^y - (1 + x)e^{-y} = 0$$

$$(1 - x)e^{2y} - (1 + x) = 0$$

$$e^{2y} = \frac{1 + x}{1 - x}$$

$$y = \frac{1}{2}\ln\left(\frac{1 + x}{1 - x}\right).$$

$\square$

7.4   Derivatives of inverse hyperbolic functions

Theorem. The derivative of $\operatorname{arsinh} x$ is $1/\sqrt{x^2 + 1}$.

Proof. Suppose that $y = \operatorname{arsinh} x$, so $x = \sinh y$. Then

$$\frac{dx}{dy} = \cosh y$$

$$= \sqrt{\sinh^2 y + 1}$$

$$= \sqrt{x^2 + 1}$$

(note that we take the positive square root since $\cosh y$ is positive).

$$\frac{dy}{dx} = \frac{1}{\sqrt{x^2 + 1}}.$$

$\square$

Theorem. The derivative of $\operatorname{arcosh} x$ is $1/\sqrt{x^2 - 1}$.

Proof. Suppose that $y = \operatorname{arcosh} x$, so $x = \cosh y$. Then

$$\frac{dx}{dy} = \sinh y$$

$$= \sqrt{\cosh^2 y - 1}$$

$$= \sqrt{x^2 - 1}$$

(we take the positive square root since $\sinh y$ is nonnegative for nonnegative y).

$$\frac{dy}{dx} = \frac{1}{\sqrt{x^2 - 1}}.$$

$\square$

Theorem. The derivative of $\operatorname{artanh} x$ is $1/(1 - x^2)$.

Proof. Suppose that $y = \operatorname{artanh} x$, so $x = \tanh y$. Then

$$\frac{dx}{dy} = \operatorname{sech}^2 y$$

$$= 1 - \tanh^2 y$$

$$= 1 - x^2$$

$$\frac{dy}{dx} = \frac{1}{1 - x^2}.$$

$\square$

8 Differential equations

8.1 First order linear differential equations

Theorem. Let $dy/dx + P(x)y = Q(x)$ be some first order linear differential equation. Then, multiplying by $\mu(x) = e^{\int P(x)\, dx}$ gives

$$\frac{d}{dx}(\mu(x)y) = Q(x)\mu(x).$$

Proof. Suppose that the result holds for some $\mu(x)$ which is positive for all $x \in \mathbb{R}$. We wish to show that $\mu(x) = e^{\int P(x)\, dx}$ is a solution for any $P(x)$ and $Q(x)$. Expanding the result by the product rule, we get

$$y \cdot \mu'(x) + \mu(x) \cdot \frac{dy}{dx} = Q(x)\mu(x).$$

Then dividing by $\mu(x)$:

$$\frac{dy}{dx} + y \cdot \frac{\mu'(x)}{\mu(x)} = Q(x).$$

And so, comparing coefficients with the original first-order linear differential equation, we get that $\mu'(x)/\mu(x) = P(x)$. Integrating both sides, we obtain $\ln|\mu(x)| = \int P(x)\, dx$. Since $\ln|\mu(x)| = \ln(\mu(x))$, we have $\mu(x) = e^{\int P(x)\, dx}$.

Note that $\int P(x)\, dx$ can be any antiderivative of $P(x)$, so we can ignore the integration constant when calculating the integrating factor. $\qquad\square$

8.2 Second order differential equations

Theorem. Suppose that we have the following homogeneous second order differential equation

$$\frac{d^2y}{dx^2} + a\frac{dy}{dx} + by = 0.$$

Now consider the roots of the auxiliary equation given by $\lambda^2 + a\lambda + b = 0$:

- If there are two distinct real roots λ_1 and λ_2, then the general solution is given by $y = Ae^{\lambda_1 x} + Be^{\lambda_2 x}$;

- If there is a single repeated real root λ, then the general solution is given by $y = Ae^{\lambda x} + Bxe^{\lambda x}$;

- If there are two complex roots $\lambda_1 = \alpha + \beta i$ and $\lambda_2 = \alpha - \beta i$ (they must be complex conjugates), then the general solution is given by $y = e^{\alpha x}(A\cos(\beta x) + B\sin(\beta x))$.

Proof. We can rewrite the differential equation using the differential operator as:

$$(D^2 + aD + b)y = 0.$$

This gives rise to an auxiliary equation $f(x) = \lambda^2 + a\lambda + b$. We consider 2 cases:

- Suppose that $f(x)$ has 2 distinct roots, λ_1 and λ_2, which may be 2 real numbers or they may be complex conjugates of one another.

  Then, factorising the differential equation

  $$(D - \lambda_1)(D - \lambda_2)(y) = 0.$$

  We let $z = (D - \lambda_2)(y)$, so $(D - \lambda_1)z = 0$.

  This gives us two first-order differential equations that we can solve:

  $$\frac{dz}{dx} - \lambda_1 z = 0 \tag{2}$$

  $$z = \frac{dy}{dx} - \lambda_2 y \tag{3}$$

  We begin with Equation 2, which can be rearranged to give

  $$\frac{dz}{dx} = \lambda_1 z$$

  Now suppose that $z = f(x)$ is a solution to this equation, so $f'(x) = \lambda_1 f(x)$. Then

  $$\frac{d}{dx}\left(\frac{f(x)}{e^{\lambda_1 x}}\right) = \frac{e^{\lambda_1 x} f'(x) - f(x) \cdot \lambda_1 e^{\lambda_1 x}}{e^{2\lambda_1 x}}$$

  $$= \frac{e^{\lambda_1 x}(\lambda_1 f(x)) - f(x) \cdot \lambda_1 e^{\lambda_1 x}}{e^{2\lambda_1 x}}$$

  $$= f(x) \cdot \lambda_1 \frac{e^{\lambda_1 x} - f(x) \cdot \lambda_1 e^{\lambda_1 x}}{e^{2\lambda_1 x}}$$

  $$= 0.$$

So $f(x)/e^{\lambda_1 x}$ is a constant function, hence $f(x)/e^{\lambda_1 x} = k$ for some $k \in \mathbb{R}$. Then $z = ke^{\lambda_1 x}$ is the general solution of Equation 2.

Now we can solve Equation 3, which is linear. We multiply by the integrating factor, which is $e^{\int -\lambda_2 \, dx} = e^{-\lambda_2 x}$:

$$e^{-\lambda_2 x} z = e^{-\lambda_2 x} \frac{dy}{dx} - \lambda_2 e^{-\lambda_2 x} y$$

$$= \frac{d}{dx}(e^{-\lambda_2 x} y)$$

$$\int e^{-\lambda_2 x} \cdot k e^{\lambda_1 x} \, dx = e^{-\lambda_2 x} y + \ell$$

$$\int k e^{(\lambda_1 - \lambda_2)x} \, dx = e^{-\lambda_2 x} y + \ell$$

$$\frac{k}{\lambda_1 - \lambda_2} e^{(\lambda_1 - \lambda_2)x} = e^{-\lambda_2 x} y - B$$

$$A e^{(\lambda_1 - \lambda_2)x} = e^{-\lambda_2 x} y - B$$

$$A e^{\lambda_1 x} = y - B e^{\lambda_2 x}$$

$$y = A e^{\lambda_1 x} + B e^{\lambda_2 x}.$$

This proves the first part of the theorem. It remains to show the result for complex solutions.

Note that if $\lambda_1 = \alpha + \beta i$ is a solution of the auxiliary equation for $\beta \neq 0$, then $\lambda_2 = \alpha - \beta i$. Substituting this into the result obtained above:

$$
\begin{aligned}
y &= C e^{(\alpha + \beta i)x} + D e^{(\alpha - \beta i)x} \\
&= C e^{\alpha x}(\cos(\beta x) + i \sin(\beta x)) + D e^{\alpha x}(\cos(-\beta x) + i \sin(-\beta x)) \\
&= C e^{\alpha x}(\cos(\beta x) + i \sin(\beta x)) + D e^{\alpha x}(\cos(\beta x) - i \sin(\beta x)) \\
&= e^{\alpha x}(C \cos(\beta x) + C i \sin(\beta x) + D \cos(\beta x) - D i \sin(\beta x)) \\
&= e^{\alpha x}((C + D) \cos(\beta x) + (C - D) \sin(\beta x)).
\end{aligned}
$$

Note that $C + D$ and $C - D$ are arbitrary constants. To demonstrate this, we will show that for any $A, B \in \mathbb{R}$, we can find C and D such that:

$$C + D = A \tag{4}$$

$$C - D = B \tag{5}$$

Adding Equation 4 and Equation 5, we get $2C = A + B \implies C = (A + B)/2$. By subtracting the equations, we similarly obtain $D = (A - B)/2$.

Then the general solution for complex conjugate roots of the auxiliary equation is

$$y = e^{\alpha x}(A \cos(\beta x) + B \sin(\beta x)).$$

- We now examine the final case, where the auxiliary equation has a single repeated real root λ. Then we can factorise the differential equation as $(D - \lambda)(D - \lambda)y = 0$. Letting $z = (D - \lambda)y$, this gives rise to two first-order differential equations:

$$\frac{dz}{dx} - \lambda z = 0 \tag{6}$$

$$z = \frac{dy}{dx} - \lambda y \tag{7}$$

Note that Equation 6 is the same as Equation 2, just with λ_1 replaced with λ. So we know that the general solution to Equation 6 is $z = Be^{\lambda x}$. We can then substitute this into Equation 7 to give a linear first-order differential equation. Multiplying through by the integrating factor, which is $e^{\int -\lambda\, dx} = e^{-\lambda x}$, we get

$$ze^{-\lambda x} = \frac{dy}{dx}e^{-\lambda x} - \lambda y e^{-\lambda x}$$

$$(Be^{\lambda x})(e^{-\lambda x}) = \frac{d}{dx}(ye^{-\lambda x})$$

$$Be^0 = \frac{d}{dx}(ye^{-\lambda x})$$

$$\int B\, dx = ye^{-\lambda x}$$

$$Bx + A = ye^{-\lambda x}$$

$$y = Ae^{\lambda x} + Bxe^{\lambda x}.$$

$\square$

Printed in Great Britain
by Amazon